Four Firefighters Killed, Trapped by Floor Collapse Brackenridge, Pennsylvania

Investigated by: J. Gordon Routley

This is Report 061 of the Major Fires Investigation Project conducted by TriData Corporation under contract EMW-90-C-3338 to the United States Fire Administration, Federal Emergency Management Agency.

FEMA

Department of Homeland Security
United States Fire Administration
National Fire Data Center

U.S. Fire Administration Fire Investigations Program

The U.S. Fire Administration (USFA) develops reports on selected major fires throughout the country. The fires usually involve multiple deaths or a large loss of property. But the primary criterion for deciding to do a report is whether it will result in significant "lessons learned." In some cases these lessons bring to light new knowledge about fire--the effect of building construction or contents, human behavior in fire, etc. In other cases, the lessons are not new but are serious enough to highlight once again, with yet another fire tragedy report. In some cases, special reports are developed to discuss events, drills, or new technologies which are of interest to the fire service.

The reports are sent to fire magazines and are distributed at National and Regional fire meetings. The International Association of Fire Chiefs assists the USFA in disseminating the findings throughout the fire service. On a continuing basis the reports are available on request from the USFA; announcements of their availability are published widely in fire journals and newsletters.

This body of work provides detailed information on the nature of the fire problem for policymakers who must decide on allocations of resources between fire and other pressing problems, and within the fire service to improve codes and code enforcement, training, public fire education, building technology, and other related areas.

The Fire Administration, which has no regulatory authority, sends an experienced fire investigator into a community after a major incident only after having conferred with the local fire authorities to insure that the assistance and presence of the USFA would be supportive and would in no way interfere with any review of the incident they are themselves conducting. The intent is not to arrive during the event or even immediately after, but rather after the dust settles, so that a complete and objective review of all the important aspects of the incident can be made. Local authorities review the USFA's report while it is in draft. The USFA investigator or team is available to local authorities should they wish to request technical assistance for their own investigation.

This report and its recommendations were developed by USFA staff and by TriData Corporation, Arlington, Virginia, its staff and consultants, who are under contract to assist the U.S. Fire Administration in carrying out the Fire Reports Program.

The USFA greatly appreciates the cooperation and information received from Chief Danny Brestensky and Assistant Chief Dennis Vrotney, Pioneer Hose Company; Chief Matthew D. Frantz, Hilltop Hose Company; Allegheny County Fire Marshal John Kaus and Assistant Fire Marshal Larry Boyle; Chief Richard Heuser, Eureka Hose Company; and Assistant Chief Terry Chambon, Highland Hose Company.

For additional copies of this report write to the U.S. Fire Administration, 16825 South Seton Avenue, Emmitsburg, Maryland 21727. The report is available on the Administration's Web site at http://www.usfa.dhs.gov/

U.S. Fire Administration
Mission Statement

As an entity of the Department of Homeland Security, the mission of the USFA is to reduce life and economic losses due to fire and related emergencies, through leadership, advocacy, coordination, and support. We serve the Nation independently, in coordination with other Federal agencies, and in partnership with fire protection and emergency service communities. With a commitment to excellence, we provide public education, training, technology, and data initiatives.

TABLE OF CONTENTS

Four Firefighters Killed, Trapped by Floor Collapse
Brackenridge, Pennsylvania
December 1991

Local Contacts: Chief Danny Brestensky
Pioneer Hose Company
122 Morgan Street
Brackenridge, Pennsylvania 15014
(412) 224-3336

John Kaus
Allegheny County Fire Marshal
Penn-Liberty Plaza
1520 Penn Avenue
Pittsburgh, Pennsylvania 15222
(412) 392-8550

Chief Matthew D. Frantz
Hilltop Hose Company
P.O. Box 214
Natrona Heights, Pennsylvania 15065

OVERVIEW

Four volunteer firefighters died when they were trapped by a partial floor collapse during a structure fire in Brackenridge, Pennsylvania, on the morning of December 20, 1991. All four were members of a mutual aid truck company that had responded to the early morning incident and were assigned to prevent fire extension from the basement to the ground floor of a 2-story building. Although they were wearing full protective clothing and using self-contained breathing apparatus (SCBA), it appears that they were overwhelmed by the severe fire conditions that erupted when a section of the ground floor collapsed into the basement. The collapse cut off their primary escape path, and the fire burned through their hoseline, leaving them without protection from the flames. (Appendix A presents the timeline of events leading up to the floor collapse.)

SUMMARY OF KEY ISSUES

Issues	Comments
Situation	Fire in enclosed room in basement. Unable to locate fire because of smoke. Smoke and heat increasing, but no visible fire.
Structure	Appeared to be heavy concrete construction. Actually thin concrete floors supported by unprotected steel.
Contents	Furniture refinishing business. Quantities of flammable finishes and solvents in basement.
Exits	One entrance/exit on each level; no alternate exits.
Structural Collapse	Floor section collapsed between interior crew and their only exit. Fire overwhelmed crew.
Rescue Attempts	Valiant rescue efforts proved unsuccessful. Unsure if missing members fell into basement or were trapped on ground floor.
Incident Command	No formal command system or personnel accountability in-place. Chief of first-due company in command of incident; Assistant chiefs assigned to basement and ground floor.
Information	No pre-fire plan and no detailed knowledge of occupancy. Clues of structural danger not recognized as fire conditions increased.
Communications	Radio system inadequate for current needs.
Response	Independent volunteer companies. Mutual aid requested on arrival and additional companies called in succession.
Weather	Extremely cold night, pre-dawn hours. Problems with frozen hydrants.
Water System	Weak supply. Extensive mutual aid and long relays needed to protect exposures.

The firefighters who died were all members of the Hilltop Hose Company. They were:

Name	Age
Michael Cielicki Burns	27
David Emmanuelson	29
Rick Frantz	23
Frank Veri, Jr.	31

The analysis of this incident provides several valuable lessons for the fire service. Unfortunately these are all revisited lessons, not new discoveries. These firefighters died in the line-of-duty, while conducting operations that appeared to be routine, and were unaware of the situation that was developing below them. They died in spite of the fact that they were experienced, they were operating with a standard approach to operational safety, and they were the object of repeated rescue attempts by highly capable comrades.

There are several factors that could have provided warning or changed the outcome of this situation. Like most accidents, this situation was the result of a number of problems that came together under the worst possible circumstances. Firefighting obviously involves inherent dangers that must be accepted by its practitioners. The important messages for the fire service are to identify risk factors in advance of an incident and to develop mechanisms to react appropriately when critical situations present themselves.

This situation bears distinct similarities to other incidents that have claimed the lives of several fire-fighters in the past. The lessons that must be derived from this incident are not a condemnation of the actions or judgment of anyone who was involved in the situation; they simply identify information that can help to prevent this type of accident from occurring in the future.

FIRE SERVICE ORGANIZATION

Brackenridge is a community of approximately 4,500 population, located approximately 25 miles northeast of Pittsburgh, next to the Allegheny River. It is protected by the Pioneer Hose Company, which operates two stations, necessitated by railroad tracks that divide the community. (See Appendix B for area map.) Each station is equipped with a pumper (E50 and E51) and an equipment truck (R50 and R51). A van is also provided at Station 50 to transport additional personnel, designated as Squad 50.

The all-volunteer company has approximately 30 active firefighting members, supported by a large contingent of auxiliary, support, and inactive members. An elected fire chief and two assistant chiefs are responsible for command functions and the line officers include a captain and several lieutenants.

Station 50 is located on Morgan Street, one block from the river front. Station 51 is also located on Morgan Street, on the opposite side of the railroad tracks. The fire building is in the downtown part of the community, at the intersection of Brackenridge Avenue and Morgan Street, within 150 feet of Station 50. (For area map see Appendix B.)

The surrounding area comprises several contiguous boroughs and townships, protected entirely by volunteer fire companies. The companies protecting neighboring Tarentum Borough include the Highland Hose Company and the Eureka Hose Company. Highland (Company 11) operates primarily as a truck company, although their ladder truck was out of service for a major rehabilitation at the time of the incident. The primary truck company tools had been transferred to Special Service Unit SS115 while the ladder truck was gone. This company also operates a pumper to provide for water supply.

Eureka Hose (Company 12) operates as the primary rescue and ambulance company for the area. In addition to a rescue-pumper (E121), this company operates a heavy rescue truck, a squirt/hose wagon, and two ambulances. Career employees maintain Advanced Life Support ambulance service during weekdays, when volunteer staffing is limited.

The Tarentum Police Department dispatches the Tarentum and Brackenridge fire companies and additional companies in surrounding townships.

Harrison Township adjoins Brackenridge to the north and east and is served by three volunteer companies, all dispatched by the Harrison Police Department. Hilltop Hose Company of Natrona Heights operates a pumper (E31), an 85-foot ladder-tower quint (T33), an equipment truck (U34), and a personnel carrier (S32). Hilltop was designated to provide truck company service for Brackenridge while Tarentum's ladder truck was out of service. The travel distance from the Hilltop station to the fire location is approximately one mile.

Citizen's Hose Company provides ambulance and engine company service to Harrison Township. Harrison Hills Hose Company provides an additional engine company, while a fourth company was recently withdrawn from service.

The communities are located near the intersection of four different counties and mutual aid is available from all surrounding jurisdictions. There are several hundred volunteer companies in the counties surrounding Pittsburgh, each with strong company identity and local community relationship. Each volunteer company operates autonomously within its designated first due area and the chief of that particular company is the ranking authority for fire and rescue operations.

The Allegheny County Fire Marshal is responsible for fire cause determination and investigations throughout the county. The fire marshal's office is organized within the county police department and has no authority for fire suppression, although most of the staff members are active within volunteer companies. Building and fire codes are adopted and enforced by the local jurisdictions. Some volunteer companies are involved in limited fire code inspection activities, where codes have been adopted by their local jurisdictions.

The fire marshal is also responsible for the county's emergency management functions. Within the scope of emergency management, Allegheny County Public Works has resources to assist the volunteer organizations with specialized equipment for major incidents. These resources are provided by county employees as an additional service of the county.

FIRE BUILDING

The fire building was located in the downtown area of Brackenridge, in an area of mixed residential and commercial occupancies, two blocks from the Allegheny River. Most of the structures in the immediate area are two stories in height, and the fire building was one of the larger structures in the immediate area, although much larger industrial buildings are located only a few blocks away. The water supply in the area is limited to between 750 and 1,000 gallons per minute.

The building was two stories in height, approximately 65 feet wide by 75 feet deep, with a full basement. (See Appendices C and D for floor plans and structure diagrams.) The vertical distance between floors was approximately 12 feet and the ground floor at the front of the building was even with the sidewalk level. The ground slopes down toward the river and at the rear of the building the basement is approximately half above grade level. A rear driveway slopes down to a roll-up garage door providing vehicle access to the basement.

At one time the building served as an automobile dealership and an interior ramp allowed vehicles to be driven to the second floor. A roll-up garage door on the front of the building provided access to the ramp. An additional roll-up door, inside the building, provided access from the bottom of the ramp into the main portion of the ground floor, which served as the showroom.

Construction details indicate that the ramp was not part of the original construction. The upper floor was originally used as a roller skating arena and became part of the automobile dealership at a later date. When the ramp was installed, the stairs to the second floor were removed, leaving the ramp as the only access to the upper level. An enclosed stairway, near the center of the building, connected the basement and the ground floor.

A set of renovation plans, dated 1980, indicated a set of double exit doors discharging from the ground floor level at the rear of the building. (This may have been the original exit discharge from the upper floor.) These doors were located in the corner, under the automobile ramp and could only be reached by walking under the ramp through an area with low overhead clearance. When a change of occupancy occurred, the exterior exhaust shaft for the basement spray booths was installed directly in front of these doors, permanently eliminating this exit.

The residential building to the rear is an architectural match for the fire building and a bridge once connected the two buildings to provide access to the owner's apartment. Two steel beams still spanned the rear driveway, connecting the ground floor of the fire building with the second floor of the building at the rear. The bridge and the door that provided access to it were removed in an earlier renovation.

There were no exterior stairs or fire escapes, and the building contained no fire alarm, detection, or sprinkler systems. The only exterior access to the ground floor was at the front of the building and the only exterior access to the basement was at the rear. An interior stairway near the center of the building linked the basement and ground levels.

The basement foundation walls were concrete, and the exterior walls above were brick over terra-cotta tile construction. A portion of the basement was separated from the main area by brick walls to serve as a boiler room, although it was no longer used in this capacity. Natural gas and electric heating units had replaced the old central furnace at some point in the building's history.

The floors were concrete; the first floor surface was finished with terrazzo and tile in different areas and the upper floor surface had been covered with wood.

OCCUPANCY

The building housed West Interior Services, a company that specialized in restoration and refinishing of furniture. The front part of the ground floor had been converted to offices and the large show-room windows had been replaced by smaller window assemblies. The ramp to the second floor remained in-place and the upper level was used for storage.

All of the company's production facilities were located in the basement. A stripping operation was located toward the rear, with three dip vats containing solvents and neutralizers that were used to remove old finishes from wood furniture. Thermostatically controlled immersion heaters were installed to keep the contents of the vats at their proper temperatures.

A 30 foot by 30 foot finishing room was constructed in the front part of the basement to provide a clean environment to apply finishes and to contain flammable vapors. Two spray booths were installed in this area and were vented to the exterior by an exhaust system. The exhaust duct carried spray vapors to the rear of the building and extended up the exterior to discharge at the roof level. An additional exhaust fan was installed at the rear exterior of the building to move the exhausted vapors up to the point of discharge.

The finishing room walls were constructed of gypsum wallboard on wood studs, framed into the building structure. Air intake filters were installed in one of the walls to provide for make-up air when the spray booth exhaust system was operating. A set of double doors allowed for large pieces of furniture to be moved into and out of the finishing room.

Shelves inside the finishing room were provided for finishing products, primarily in one gallon and five gallon metal containers, including lacquers and lacquer thinners. A considerable quantity of aerosol containers of touch-up and special finish materials was kept on a set of shelves in one corner of the room. Other materials used in the finishing process were stored on shelves and cabinets in the room.

A self-closing rag container was provided in the room. An area just outside the double doors was provided for 35 and 55 gallon drums of flammable liquids, which were connected to a grounding system. Liquids from these containers were dispensed into smaller containers for use in the finishing room and in other areas. Several other drums containing flammable and non-flammable products were stored in the basement and in an exterior storage area at the bottom of the rear driveway.

The remainder of the basement was used for woodworking and storage. A rack in one corner was used to store lumber for repair jobs. The area between the finishing room and the roll-up door was used as a shipping and receiving area where finished jobs were staged for shipment and newly arrived pieces were unloaded. The entire basement was cluttered with woodworking equipment, work in progress, materials in transit, and miscellaneous storage.

THE FIRE

The fire originated in the basement finishing room, which is directly below the front entrance. The investigation indicates that the cause was most likely an accumulation of overspray residue that was ignited by the electric heating unit. The alternative possibility of spontaneous ignition of chemicals used in the room could not be eliminated. The actual time of ignition is believed to have been as long as several hours before discovery. Investigators believe that the fire may have smoldered for several hours and may have gone through repeated stages of flaming and smoldering combustion, limited by the ventilation that was available to sustain flaming combustion.

Between midnight and 0100 hours neighbors had reported a strong odor of smoke in the area and members of the Pioneer Hose Company, who were at the station, had looked for a source, concluding that it was coming from a wood stove. The temperature was approximately 10 degrees Fahrenheit, with no wind, and many wood stoves were in use in the neighborhood. One witness later reported that visible smoke had been coming from either the chimney or the basement exhaust duct, but this was not observed by the firefighters who checked the area.

Around 0200 hours another call was received by the communications center and a police car was sent to check to area, but the officer found nothing unusual.

The fire was discovered by the first employee reporting for work, on the morning of December 20, 1991. The employee discovered smoke immediately upon opening the front door, then went to the rear and observed smoke in the basement. The employee knocked on the door of the residential building to the rear and asked a resident to call the fire department. The Tarentum Police Dispatch Center received the call at 0545 hours, reporting smoke in the building.

INITIAL RESPONSE

The Tarentum dispatcher activated the tones to notify Pioneer members of the alarm at 0546 hours. The Pioneer Chief, who lives less than a block from the scene, was first to arrive and reported "smoke showing" at 0547. He immediately requested the response of Hilltop Hose Company for their ladder truck. This request was relayed from the Tarentum dispatcher to the Harrison dispatcher and the company's tones were activated at 0549.

One of the first Pioneer firefighters to arrive at the station dressed in full protective clothing picked up an SCBA and a forcible entry tool, and responded to the front of the fire building on foot. At this time light smoke was coming from the front door, which had been opened by the employee. As the firefighter stopped inside the doorway to don the SCBA facepiece, he noted that the floor was hot to touch with a gloved hand and through the kneepads of his turnout pants. He concluded that the fire must be in the basement and went around to meet the crew that was arriving with Engine 50.

The fire chief directed Engine 50 to lay a 5-inch supply line from the hydrant next to the station to the fire building. This was done as soon as sufficient members had arrived to respond with an engine company crew. Engine 51 was responding from the other station and arrived slightly after Engine 50. The fire chief directed Engine 51 to lay an additional 5-inch supply line from a hydrant in their direction of approach. As Engine 51 arrived, one of the crew members noted heavy smoke coming from the chimney at the rear and suspected that the problem could be a chimney fire. Additional members responded on the rescue trucks from each station at 0553 hours.

INTERIOR OPERATIONS

Engine 50's crew opened the rear door and extended a 2-inch attack line down the five steps into the basement. They encountered moderate smoke and heat, but no visible fire. They were able to walk upright and navigate with handlights, although their vision was extremely limited by the smoke. The line was advanced approximately 50 feet into the basement as they worked their way around the stripping vats and furniture.

Hilltop's Truck 33 was en route with a seven member crew and was requesting approach directions from the Pioneer Chief by 0555 hours. Truck 33 was directed to the front of the building and was on the scene by 0559 hours. The crew included the Hilltop Assistant Chief and the company's second lieutenant, who were brothers, and five additional members. The Hilltop Chief responded on Engine 31 with a six member crew approximately three minutes behind the ladder truck.

Pioneer's two Assistant Chiefs (C5 and C53) also arrived at the front of the building and conferred with their chief. One of the assistant chiefs (C53) assumed responsibility for interior operations on the ground floor and made an entry, with the building owner, to unlock doors. The owner, who had responded from this home, was a former volunteer firefighter and reported that he was able to enter and briefly walk around on the ground floor at that time (between 0555 and 0600), although the smoke was a strong respiratory irritant.

On arrival, Truck 33 was requested to provide an interior entry team for the ground floor. The lieutenant and three firefighters donned the four SCBAs that are carried on the truck and reported to C53 at the front door. Before entering, the lieutenant and the Hilltop Assistant Chief switched their portable radios to a tactical radio channel because of the heavy traffic on the main channel. This placed them on a separate channel from the other units on the scene of the fire.

The entry team advanced a 2-inch attack line to the interior, accompanied by two crew members from E51, to search for signs of fire. The remaining members of Truck 33's crew remained outside and were joined by Engine 31's crew. The Pioneer Chief assigned them to perform ventilation by breaking a window on the second floor at the front of the building. At this point (approximately 0605 hours) arriving members noted a considerable amount of smoke coming from all openings in the building, although it did not appear to be particularly hot or thick.

The other Pioneer Assistant Chief (C5) assumed responsibility for operations at the rear of the fire building. When he surveyed conditions from the basement door, he was concerned with the increasing smoke conditions and the fact that the attack line crew could not locate the fire. He was also concerned with crew accountability and had some difficulty making contact with Captain 50, who was inside the basement trying to open the overhead door. Attempts to open the roll-up door for ventilation were unsuccessful. He directed the crew to back the line out of the basement until ventilation could be accomplished.

MUTUAL AID

The Eureka and Highland Hose Companies were requested for mutual aid at 0600 and 0601 respectively. Eureka's Engine 121 was requested to provide a water supply for Engine 50, when it was found that their hydrant was inoperative. A medic unit had already been requested from Eureka to stand by at the scene. The second request brought five members on Engine 121, followed by four members on the rescue truck and Eureka's Chief.

Engine 121 was assigned to extend Engine 50's line to another hydrant. This hydrant was also found to be inoperative, and the line had to be extended 400 feet to a third hydrant to finally obtain a water supply. This was completed at 0617 hours, 30 minutes after the first unit arrived at the scene.

Engine 51 also had problems with their supply line and had to shut down their hydrant to reconnect the 1/4-turn coupling that disengaged from the pumper when the line was charged. Although both pumps had problems with their supply lines, they were able to charge the attack lines with tank water and both had corrected the problems before any of the attack line crews encountered the fire.

Highland's Special Service 115 was requested to provide additional truck company capability and responded with seven members. At 0611 hours C5 requested the SS115 crew to assist with forcible entry on the roll-up door at the rear of the building. (Appendix E lists the fire companies involved in this incident.)

CHANGING CONDITIONS

By the time the Highland crew was requested to cut the door, the smoke conditions in the basement had begun to change significantly. C5 advised the fire chief (C50) at 0611 hours that the fire in the basement was becoming more serious and that the basement looked as if it might "light up." He was planning to cut the roll-up door to accomplish ventilation of the basement.

Two minutes later, at 0613, C5 advised C50 that he could hear the fire building up in the basement. He had entered the basement and could hear the sounds of crackling and popping, indicating that a significant fire was burning somewhere in the basement. Heavy smoke was now coming from basement openings and ventilation was needed. Crackling noises and muffled explosions could be heard, and the smoke was becoming hotter and darker. The smoke movement suggested that backdraft conditions could be developing within the basement.

Members from SS115, who were bringing a saw to the rear to begin cutting the door, noted smoke pushing from the joint at the intersection of the front wall and the sidewalk. Crew members who crawled into the basement, after cutting the bottom panel out of the roll-up door, encountered elevated temperatures when they attempted to stand up inside.

At 0618 C5 again advised the chief of the need for ventilation and reported that he was reassigning the attack crew to take out the basement windows on the Morgan Street side of the building. The first four windows opened into the boiler room and, when they were broken, only moderate smoke was released. The fifth window was the only one that opened into the main part of the basement and when it was broken a heavy volume of hot black smoke was released.

At the front of the building, at this time, the volume of smoke was increasing, but it was still relatively light in color and lazy in movement. There was no indication of visible fire. The Pioneer Chief discussed the need for vertical ventilation with the Hilltop Chief and instead requested that Hilltop's members break out additional windows on the upper floor at the front and side. Hilltop's members split into two crews and positioned ground ladders on the front and side of the building to accomplish this task. Electrical power lines required extra caution in positioning of the ground ladders.

The crew members from Hilltop T33 (the four men who are to die in this fire) had advanced their line in through the front door to the open area near the top of the basement stairs, and the Pioneer Assistant Chief (C53) went outside to obtain the key to the stairway door from the business owner. He returned and unlocked the door, but it was hot to the touch and smoke was issuing from under and around it. He directed the crew to hold their position in the open area and to look out for vertical extension of the fire, while he went back outside to confer with the fire chief.

With the heavy volume of smoke coming from the basement window, C5 determined that the best direction of attack would probably be via the basement stairs, using a positive pressure ventilation fan to keep the stairs clear and to push the smoke and heat out through the basement windows and the overhead door. One of the crew members was sent to determine the location of the stairs and the ability to advance a line by that route.

FIRE ATTACK

At approximately 0620 hours, the Eureka Rescue crew and the members who had completed the supply line for Engine 50 walked up to the rear of the building and encountered the SS115 crew still attempting to cut through the rear roll-up door. Some of the members began to assist with cutting the door, while others obtained a 1-3/4-inch attack line from E50. The attack line was advanced under the door and fire was immediately visible along the wall to the left of the doorway and toward the corner where the finishing room was located. The line was advanced into the basement as the crew attempted to knock down the visible fire. At approximately 0623 hours, this was the first sign of visible fire and the first application of water on the fire.

The fire conditions at this point suggest that the fire had finally broken out of the finishing room, possibly due to a failure of its enclosure wall. The Eureka members worked their way into the basement, but could not knock down the heavy volume of fire in the far corner with the flow from their 1-3/4-inch line. The Eureka Chief directed the crew to back out and sent other members to obtain a large line and a portable master stream to place in the doorway. Two attempts to contact the Pioneer Fire Chief by radio were unsuccessful.

FLOOR COLLAPSE

At the front door, the assistant chief (C53) was just coming out as he passed the firefighter who was entering to locate the stairs. The firefighter took one step inside the front door and began to sink as the floor began to collapse into the basement. He was pulled to safety by the assistant chief as the floor dropped away. In an instant the interior erupted in a rush of smoke, followed by a ball of fire, blowing out the doorway and shattering the ground floor front windows.

The eruption pushed the firefighters into the street and knocked several others off their feet. As they looked back toward the doorway they could see the 2-inch hoseline rupture as it was enveloped by fire. Heavy fire continued to pour from the basement and out through the front openings at street level.

As the floor collapsed there was only enough time for someone to shout "Get 'em out!" over the radio. There was no contact with the four Truck 33 crew members. Some members realized immediately that firefighters could be trapped inside and additional hoselines were quickly advanced and operated into the doorway, but they could not suppress the fire. There was no possibility to advance lines into the building – the floor was completely gone across the entire front of the building.

There was no rear exit on the first floor, so the only way the T33 members could have gotten out of the building would have been to go down the stairs and out the door at the rear of the basement.

There was some initial confusion over the number of members who were missing and their specific identities. Hilltop members knew that they had four crew members inside, but only three were immediately identified. The identity of the fourth crew member had to be determined by reconstructing who had seen whom on the truck and donning SCBA prior to entering the building. Pioneer members had been in and out of the building and there was no system in-place to rapidly determine who had responded or where they were on the incident scene at the time of the collapse. The Pioneer officers had to account for all of their members to confirm that all were out of the building.

The word spread rapidly that at least three members were missing and were believed to have fallen into the basement inferno. It was approximately 30 minutes before the actual number and their identities were confirmed.

While the situation was obvious to everyone at the front of the building, the crews operating at the side and rear were unaware of the critical situation for several minutes. While the ground shook and the building erupted in flames at the front, the only change that was evident in the other areas was a brief "push" of smoke followed by a significant increase in smoke and heat conditions. The Eureka crew that was backing out with the 1-3/4-inch line was still in the basement and its members were unaware of the collapse. The members who were working on the roll-up door now had hot heavy black smoke coming out through their opening.

RESCUE ATTEMPTS

An estimated ten minutes elapsed before it was known to the crews at the rear that a collapse had occurred and that members were believed to have fallen into the basement. The Eureka crew organized an entry team to attempt a rescue while the portable deluge was used to attack the fire in the basement. The entry team advanced a hoseline through the basement to the point where they encountered collapse steel and concrete, but could find no sign of the missing members. Based on the report that the Hilltop crew had last been seen near the top of the stairs, they advanced the line up to the stairway landing. A member worked his way to the top of the stairs and reached out to search for victims, but the visibility was almost zero and there was no sign of the missing crew.

Two more rescue attempts were made by the Eureka crew and one by a Pioneer crew, in each case looking for the missing members in the basement rubble and from the top of the stairs. The upper floors of the building were becoming heavily involved in fire and the members could feel the vibrations as sections of the upper floor and roof collapsed.

The final attempt was made more than an hour after the floor collapse occurred and involved a Eureka rescue team using lifelines. From the top of the stairway, two members worked their way out to the edge of the collapse area and looked for signs of the missing members, but still found no sign of them. To reach the stairs they had to wade through water above the tops of their boots, feeling their way to avoid obstructions. As they worked their way out, they discovered the stripper solvent tanks that were overflowing into the flooded basement and several drums that were overturned or ruptured. The entire crew had to be decontaminated in the freezing temperatures and were then transported back to their station. After this attempt, all rescue efforts were suspended until the fire could be brought under control.

CONTINUED FIRE SUPPRESSION

After the collapse, the upper floors of the fire building were rapidly involved in fire. The ramp opening allowed the fire to fully involve the storage area on the second floor and the wood roof structure. Most of the first floor was also consumed by the flames. Exposure protection soon became critical, particularly to the wood frame residence on "Side 4." Additional mutual aid companies were summoned and master streams were placed into operation to confine the fire.

To supply the master streams, large diameter hoselines were stretched to the river front, where a pumper was placed to draft, and to hydrants several blocks away. Water was obtained from the Tarentum water system and from the private system at a steel plant within the limits of Brackenridge. As the building was consumed some of the exterior brick walls began to fail, bringing down the electrical power lines that ran along the front and side of the structure. The resulting power failure disabled the major pumps supplying the Brackenridge water system and more mutual aid companies had to be called to further supplement the water supply. A tanker shuttle was also used to supplement the water supply for a period of time.

The fire consumed the entire roof and second floor levels, continuing to expose neighboring structures until mid-morning. As the structure collapsed inward, elevated master streams were able to bring the flames under control. In the crowded area where the fire occurred, there was a significant risk of fire spread to other structures around the burning structure and most of the fire suppression effort was directed to exposure protection.

BODY RECOVERY

After the last rescue attempt, it was finally recognized that there was no hope of finding the missing firefighters alive. They had either fallen into the roaring inferno of the basement or they were still somewhere on the part of the first floor that remained standing, but the entire building had been heavily involved in fire for more than two hours. Heavy equipment was needed to move rubble to search for the bodies.

The Allegheny County Fire Marshal assumed responsibility for the investigation and County Public Works equipment was brought in to move the debris. A "gradall" was used to pull out the portions of the front wall that were still standing and to dig into the rubble. It was only at this point that an assessment of the floor collapse could be made and it could be seen that only the very front portion had fallen into the basement.

The bodies were located almost six hours after the collapse occurred. All four bodies were found together, approximately 35 feet inside the front door, in the same area where they had last been in contact with the assistant chief. The floor in this area did not collapse, but the bodies were almost covered by debris that had fallen as the building burned. One member still held the nozzle and another had an axe; a portable radio and a handlight were found with the bodies.

It appears that the four were immediately overwhelmed by the fire erupting from the basement, and they had no opportunity to take any action. Most of their protective equipment was destroyed, but examination of the recovered components and autopsy reports indicate that all were wearing full protective clothing and using their SCBAs when they were overcome. There was no indication of any inhalation of smoke or superheated gases.

As an example of the degree of destruction, all that remained of an SCBA cylinder was a blob of melted aluminum attached to the steel cylinder valve. The glass fiber reinforcement filaments maintained the shape of the cylinder. Although two of the SCBAs had personal alert safety system (PASS) units, no sound of PASS alarms was heard by anyone outside or during the rescue attempts. The only sound that was heard was one member who reported hearing a low pressure alarm bell on an SCBA, minutes after the collapse.

CONSTRUCTION DETAILS

The floors and roof of the fire building were supported by a frame of unprotected steel members. The side walls of the building were load bearing, as indicated by pilasters that were visible on the Morgan Street side of the building. (This detail was not evident on the opposite wall, which did not have facing brick, since it was not a street face.) Steel beams spanned the width of the building at four locations, dividing the structure into 5 bays of approximately 15 foot depth.

The massive roof girders (30 inches x 12 inches) appear to have been unsupported across the entire 65 foot width of the building. The roof girders were steel I-sections, fabricated from plate and angle components that were riveted to form the desired shapes, reflecting the structural steel technology of the period. The wood roof imposed a lighter load than the combination of cars and concrete on the floors, making the broad span feasible.

The floors were supported by steel beams and columns. The beams and columns were arranged in four major support assemblies, spanning the width of the building at 15 foot intervals. The first support (closest to the front of the building) was made up of two beams, each spanning approximately 32 feet, supported by a single center column. The three remaining supports each had two columns, with a 32 foot center span and outer spans of approximately 15 feet on each side. The outer span beams were considerably smaller than the middle span beams, due to their shorter span and reduced load. The outer ends of the beams were solidly bricked into the side walls, providing rigid anchorage at each end, and rigidly connected to the intermediate columns with bolts and brackets.

The majority of the structural steel was exposed. The only evidence of fire resistant protection for the steel frame was on the beams supporting the ground floor and provided an extra measure of fire resistance for these particular members only. The webs of the protected members had been encased in concrete, leaving the flanges and column connectors exposed. The concrete encasement appears to have been done after the original construction and the reason for it is not known. There were no provisions to anchor the concrete to the steel and the encasement on the first support fell off at some point in the structural collapse. The columns and the remainder of the beams were unprotected, although some were boxed-in by the construction of interior partition walls.

The most critical detail was the construction of the floor assemblies. The floors were supported by unprotected steel joists, approximately 14 feet 9 inches long, spaced on approximately 12 inch centers. The joists spanned the distance between the beams, and between the beams and the front and rear walls. They were fabricated of light gauge steel plates, channels, and angles, approximately 1/8-inch thick, spot welded to form the desired shapes. The fabricated joists were I-sections, 10 inches tall with 2-1/2-inch wide flanges. These lightweight steel joists were extremely vulnerable to the heat of a fire.

The anchorages of some of the joists at the front of the building were severely corroded, particularly in the area under the door leading to the vehicle ramp. These particular joists were so severely corroded that they may not have been able to support a routine load.

The concrete floor slabs were only 2-1/4-inches thick, with approximately 3/8-inch thick topping of tile or terrazzo. The only reinforcing in the concrete appears to have been a steel mesh that was used to support the bottom of the slab when it was poured.

STRUCTURAL COLLAPSE

The primary structural collapse was caused by the failure of the steel joists supporting the floor slab above the finishing room. All of the joists between the front wall and the first set of support beams failed and dropped the concrete slab into the basement. Examination of the joists found in the rubble indicates that all were severely warped and twisted from heat exposure. The slab broke near the mid-point of the span and some sections were left hanging vertically from the beams. This created an opening, directly over the fire, extending 15 feet in from the front for the full width of the building.

The middle column supporting the first set of beams was also buckled, although this appears to have been a secondary failure. The buckling of this column caused a secondary slab failure between the first and second sets of beams. The slab in the second bay sloped down toward the opening, with a low point at the center column, but did not drop into the basement. The remainder of the ground floor was not compromised.

The entire roof and most of the second floor collapsed into the rubble during the fire. The initial structure collapse involved only the one section of the first floor slab. The center column failed later, after severe fire exposure. The remaining beams and columns in the basement were not compromised, although most of the joists were warped and twisted.

Unprotected steel is particularly vulnerable to fire exposure. Structural steel loses most of its strength between 1,000 degrees and 1,400 degrees Fahrenheit and the endurance of a particular member is directly related to its mass, the load on the member, and the temperature of the fire environment.

The mass determines the time it will take for a member to be heated to its failure temperature when exposed to a fire. Light gauge members, such as the floor joists in this building, may be vulnerable to collapse with as little as three to five minutes of direct exposure to a fire, while heavier members may take 10 to 20 minutes to reach their failure temperature. Very heavy steel members may survive extended exposure to a fire environment.

The method of attachment of the joists is also very significant. Unrestrained members tend to elongate, while rigidly restrained members will warp or twist. The light steel joists were fully restrained at the outer end and were supported by a much heavier beam at the interior supports. When heated, they would tend to warp and twist, but they would probably remain connected to the floor slab and to the end supports. An assembly, such as the combined system of floor joists supporting the concrete floor slab, will tend to act together and will have a longer fire endurance than the individual members until the whole assembly reaches a point of failure.

The steel joists above the finishing room were directly exposed to fire for a long period of time. The hot floor condition was noted by the first firefighters entering the building, approximately 35 minutes before the collapse occurred, and it is possible that the fire had been burning in the finishing room for an extended period of time before being discovered. The individual joists were probably heated to the point of failure well before the collapse occurred; probably before the fire department arrived. They did not collapse immediately because they were supported by the rigidity of the combined joists and slab assembly and the manner of support at the basement wall.

The failure may have been initiated by any one of several causes. The concrete slab may have been heated enough to fracture from internal stresses or a minor backdraft in the basement could have caused it to lift slightly and then drop back. The corroded joist ends may have failed first, resulting in a "domino effect" collapse across the front of the structure, or the first movement of the buckling column could have fractured the slab and triggered a total failure of the floor section. Any of these sources could have initiated the failure. The critical point is that the failure could have occurred at any time, before or after, the arrival of the fire department.

LESSONS LEARNED

1. **An effective pre-fire planning program should cover all major structures in the community, even those that appear to be of fire resistive construction.**

    The critical details of construction that made this structure vulnerable to collapse should be recognizable to individuals who have studied the hazards of building construction. The fact that unprotected steel construction is extremely vulnerable to rapid failure under fire exposure should be clearly understood by all firefighters. The recognition of this type of construction is much more difficult while a fire is in progress than during inspection or pre-fire planning visits.

    Most individuals who were familiar with the building had the impression that it was heavy-duty solid construction, particularly since it had been used as an auto dealership and supported the weight of cars. Witnesses referred to the weight of 14-inch thick concrete floors in describing the soundness of the structure. Even the building owner believed that it was fire resistive construction and employees at the scene were quoted as referring to the "heavy duty" floor construction.

    For code purposes, this building would have been classified as "unprotected non-combustible" construction. It had no fire resistance and should have been considered vulnerable to collapse from any significant fire exposure. (It should also be noted that automobiles are considered to be a relatively light load for design purposes.)

    The extremely limited access/exit conditions should also have been noted in pre-fire planning. The only access to the ground floor was at the front, and the only access to the basement was at the rear. Access to the second floor was limited to the ramp. The firefighters on the ground floor had no alternate exit path. These are critical factors that should be incorporated into a plan that can be used by the incident commander to make critical strategic decisions during a fire.

    Due to the fact that the building was so close to the fire station and was one of the major structures in the community, several members were familiar with it in a general sense. Unfortunately, it had not been pre-fire planned and evaluated in a manner that identified its inherent weaknesses. This is a case where working from perceptions resulted in a very false sense of security.

    An effective pre-fire planning program involves critically examining buildings to identify fire risks and protection factors. This information must be documented and recorded in a systematic manner, so that it can be used for training and to support the development of a safe strategic plan during a real incident.

    The investigation also revealed that the fire building had several possible violations of the Fire Prevention Code adopted by the Borough of Brackenridge in 1981. The enforcing authority for this code would have been the borough; and it is not clear if this responsibility had been delegated to the fire department, either formally or informally. One of the objectives of a fire code is to identify and cause correction of situations that could pose a danger to firefighters during fire suppression operations. Fire code enforcement activities and pre-fire planning should be coordinated.

2. **The need for standard operating procedures (SOP) for incident management is particularly great in areas where there are numerous autonomous fire companies.**

While some of the basic elements of an incident management system were employed at this incident, there was not a clearly defined and documented system to develop a strategic plan and to effectively manage resources, particularly where several different companies are involved in the operation.

An area that encompasses numerous autonomous jurisdictional units has a particular need for SOPs to manage incidents. The incident management system must be applied in a consistent manner to effectively integrate the efforts of mutual aid companies and, particularly, to provide for operational safety.

Three essential elements of an incident management system are command organization, personnel accountability, and information management.

Command Organization – The role of the Pioneer Fire Chief as the incident commander and the assignment of the two assistant chiefs as section or division officers followed an established plan for that department and was considered a normal procedure for the companies that respond with them. Their roles were determined among themselves, however, and their assignments were not clearly identifiable, particularly to other companies arriving at the scene.

There was no standard terminology to define their roles or to identify their assignments, visibly or over the radio. Part of an effective incident management system is the ability to clearly identify who is responsible for a particular aspect of the incident.

There was no structured plan to utilize command officers from assisting companies in the incident command organization or to delegate identified roles to other qualified individuals.

Personnel Accountability – Many volunteer departments have difficulty accounting for who is at the scene, where they fit into the organization, and what they are assigned to be doing. Every member at the scene should be assigned to a particular function within a supervisory chain of command. Supervisors should be able to immediately account for the location and function of every individual or unit within their span of control. Freelance operations are extremely dangerous.

Information Management – A complex incident involves the processing of a large amount of information under stressful conditions. The incident commander must be able to gather and process information in a manner that supports the development of a plan and the continuing management of the incident. This particularly involves separating critical information from distractions that can prevent the incident commander from identifying key factors and making important decisions. The incident management system must include a component to record and process information, which could range in complexity from a clipboard to record critical information, to an aide assigned to manage and record information for the incident commander, to a designated planning function with maps, pre-fire plans, and similar capabilities, possibly including on-scene computer capabilities.

3. **Fireground information must be effectively communicated and processed to formulate a risk assessment and attack plan.**

This is a particular example of a situation where the nature and extent of the fire were not clearly identified, and the incident commander had great difficulty developing a plan to deal with the situation. Units were on the scene for over 30 minutes, with an obvious working fire somewhere in the building, before any water was applied to the fire. While several clues were present to suggest the location and growing magnitude of the fire, the information did not come back to the incident commander in a manner that supported an effective risk assessment or the development of an operational plan.

These same factors have been identified in several previous incidents that resulted in multiple firefighter deaths. In most of these cases, crews were working in areas where conditions appeared to be "routine" and non-threatening, unaware of critical factors that were occurring around them.

Analysis of an incident can make critical observations and factors obvious, after the fact. In this incident several factors were noted by different individuals (*hot floor, increasing smoke and heat in the basement, smoke pushing out between the wall and sidewalk, crackling and popping noises*). The important lesson is to be able to make these observations, identify their critical nature, communicate the information in an effective manner, and process the information in a manner that causes the hazards to be recognized. This type of information must be communicated and must be used to develop and revise the operational plan for the incident.

The "20 minute rule" is often used as a guideline in making an assessment of structural conditions. This "rule" is based on the body of experience which suggests that an "ordinary construction" building (non-fire resistive) should be considered vulnerable to structural collapse after 20 minutes of fire involvement. The rule is based on generalized experience and has many exceptions. *In the case of unprotected steel construction this could be an overly generous time allowance, while it would be extremely conservative for fire resistive construction.*

4. **An effective communication system is an essential tool of a modern fire department.**

The inadequacies of the existing communications system are evident in the analysis of this incident. Several communities in this part of Allegheny County share a common primary radio frequency. Both the Tarentum and Harrison dispatch bases and several other communities use this one frequency to dispatch volunteer fire and ambulance companies. The same frequency is used for dispatch tones, communications between individual units and the dispatcher, communications between the incident commander and the dispatcher, and on-scene tactical communications. It is not uncommon for a base station to override on-scene communications at an incident. Only one alternate tactical channel is available and it is not commonly used. In addition, there are no mutual aid or tactical channels that will accommodate units from all four of the intersecting counties. There is an obvious need for designated and coordinated tactical channels that can be used by all of the companies responding to an incident.

At this incident, the Hilltop interior crew switched to the tactical channel because of the heavy traffic on the primary channel. This restricted their ability to communicate with anyone except their own chief officers, who were engaged in other activities, and cut them off from radio communication with the incident commander or the officer responsible for their assigned area.

They were also unaware of the reports coming from anyone else at the incident scene that could have given an indication of the situation that was developing. If an evacuation order had been given, they would have had to depend on someone to repeat the order over the tactical channel.

It is extremely important to maintain communications with all units on the fireground, particularly units assigned to interior positions. This may require multiple channels and the assignment of units to different channels must be coordinated as part of the incident management system. All tactical communications must be monitored by designated individuals in the command structure.

The dispatchers at both Tarentum and Harrison are primarily police dispatchers and the fire department function is a secondary role (although some of the individual dispatchers are volunteer firefighters and are well oriented toward this role). Each location operates with a single person on-duty who must handle both police and fire radio channels and telephone communications for multiple jurisdictions. While this arrangement may work adequately in routine situations, it does not provide the level of support that is needed to effectively manage a major incident.

There is an obvious need for a total evaluation of the communications capability for the fire service in this area.

CONCLUSION

It is a sad reality that the four volunteer firefighters who died at this incident were operating at full compliance with written SOPs and safety guidelines adopted by their company. Their performance appears to have been fully reasonable and standard under the circumstances. They were operating together, as a crew, with a company officer, under an assigned command officer. They were using full protective clothing and SCBAs. They had a portable radio, lights, tools, and a charged hoseline. They died "by the book," in spite of the exemplary efforts of rescue teams who risked their own lives trying to save their fellow firefighters.

The lessons that should come from this incident should not reflect negatively on any individual. They are important lessons for the "system" – lessons that can help the fire service avoid future tragedies.

APPENDIX A

Timeline of Events

Entries in plain text are from radio tape with recorded times indicated. Entries in italics are from individual recollections of events, linked to time sequence as approximated by the individuals involved.

0544   Employee arrives for work, discovers smoke in building.

0545   Fire reported by employee to Tarentum Police dispatcher.

0546   Pioneer Hose Company (E50) dispatched by Tarentum Base.

0547   Pioneer Fire Chief (C50) arrives, reports smoke showing.

0548   C50 requests mutual aid for Truck 33, Hilltop Hose Company.

0549   Hilltop Hose Company dispatched by Harrison Base.
        Firefighter responds to scene on foot from station; stops in front doorway of fire building to adjust SCBA; discovers concrete floor is hot to touch through gloved hand; concludes that fire is in basement and proceeds to rear of building to meet E50.

0551   C50 directs E50 to lay 5-inch line from hydrant next to fire station.
        E50 lays 5-inch supply line from hydrant next to station to position on Morgan Street near rear driveway of fire building. Crew advances 2-inch attack line to basement via rear door and stairs. Moderate smoke and heat condition encountered — no fire evident.

0551   E51 responding.

0552   C50 directs E51 to lay line from hydrant at ball field.
        E51 lays supply line from hydrant one block away to position on Morgan Street side near front of fire building. Crew notes heavy smoke from chimney at rear.

0553   R50 and R51 responding.

0554   C50 requests ambulance to stand by at scene.

0554   Eureka Hose Company ambulance dispatched by Tarentum Base.

0555   T33 requesting orders; directed to front of building by C50.

0556   E50 calls for supply line to be charged.
        Firefighter finds hydrant is inoperative; possibly frozen.

0557   C50 requests report from Captain 50. Captain 50 is with crew attempting to locate fire in basement.
        Crew has advanced line approximately 50 feet into basement, but still cannot locate visible fire. Attempted radio contact is unsuccessful.

0558   E31 responding.

Appendix A (continued)

0558   C50 advised of problem with E50's hydrant.
Pioneer Assistant Chief (C53) has arrived and met with C50 at front of building--assumes responsibility for ground floor interior. Initial entry is made with building owner to unlock interior office doors. Ground floor condition reported as light smoke with no evidence of fire.

0559   C53 requests entry crew from T33.
Four crew members from T33 don SCBA and report to C53 at front door. Crew advances 2-inch line from E51 to interior.

0600   C50 requests mutual aid for E121 to supply water for E50.

0600   Eureka Hose Company E121 dispatched by Tarentum Base.

0601   E31 requesting orders.

0601   C50 requests mutual aid for SS115 to respond for additional truck company assistance.

0601   Highland Hose Company SS115 dispatched by Tarentum Base.

0602   Engine 51 has difficulty connecting supply line.
Coupling disconnects from engine when line is charged. Line is shut down and 1/4-turn coupling is reconnected and secured.

0603   E121 responding.
Assistant chief (C5) has arrived and met with C50 at front of building--assumes responsibility for rear sector.

0603   C5 attempting to reach Captain 50 – unsuccessful.

0604   E31 on scene.
C50 directs E31 and remaining crew members of T33 to vent second floor left front windows. Ground ladder is raised and moderate smoke is encountered when window is broken.

0605   SS115 requesting orders.

0606   C50 directs C12A (Engine 121) to hydrant.
E121 extends E50's supply line to designated hydrant. This hydrant is also found to be inoperative. Line is extended to next hydrant and E121 hooks up to supply pumped water to E50.

0610   C50 requests Pioneer Auxiliaries to respond to station.

0611   C5 requests SS115 to come to rear to open roll-up door. Advises C50 that fire conditions in basement are becoming serious and ventilation is needed.
C5 has made an assessment of conditions at the rear and is concerned with deteriorating conditions in basement. Smoke is becoming thick and hot, but fire has still not been located. Attack line is backed out while personnel attempt to open large roll-up door for ventilation.

0613   C5 advised C50 that he can hear fire in basement building up.
C5 can hear fire crackling and small explosions (possibly aerosol cans exploding). Smoke conditions becoming dark and heavy.

0613   C50 requests Brackenridge Maintenance Department because of icing conditions.

0614   C50 requests electric power company to respond.

Appendix A (continued)

0615   C50 requests gas company to respond.

0616   C30 asks C50 if he wants to have roof opened for ventilation.
Hilltop members have raised 35-foot ladder to roof to prepare for vertical ventilation.

0617   C12 advises E50 that water is on the way.

0618   C50 directs C30 to hold up on opening roof.
C50 directs Hilltop members to break additional second floor windows for ventilation. Crews use ground ladders to break top right windows on Brackenridge Avenue and Morgan Street sides of building. Smoke is heavier, but not severe.

0618   C5 advises C50 that he is venting basement windows on Morgan Street side of building.
Pioneer members who had been working in basement are reassigned to break basement windows on the Morgan Street side. The first four windows open into boiler room and only moderate smoke is released. The fifth window opens directly to the basement and heavy, hot, black smoke is released.

0619   C50 requests Citizen's Hose Company to stand by at Station 50.

0620   Citizen's Hose Company dispatched by Harrison Base.

0623   C12 (Eureka Chief) attempts to contact C50--no contact.
After completing supply line hook up to E50, members of Eureka Hose Company assist SS115 members attempting to open rear roll-up door. Bottom panels are cut out and 1-3/4-inch line is advanced to basement interior. Crew encounters heavy fire in direction of finishing room and moderate fire along wall to left of entry point. Line is advanced approximately 50 feet, but crews are unable to knock down heavy fire.

0624   C12 attempts to contact C50--no contact.
C12 orders crew to back out of basement. Company members are directed to extend portable deluge set to rear door to attempt to knock down large volume of fire. C53 has obtained key to basement stairs from building owner and unlocked door. Door is hot and smoke is pushing under door. Decision is made to leave door closed. Attack line crew is directed to stay in the open area and guard for vertical extension. C53 leaves to confer with C50 outside.

0626   "Come-on! Get Out!"--(unidentified voice on radio).
Front section of ground floor collapses into basement. Heavy fire blows out front door and windows. Members in front doorway are blown out into street.

0626   C5 attempts to contact 11A in rear of building.

0626   "Get a 2-1/2!"--(unidentified voice on radio).

0628   C5 requests master stream in front door.

0629   C5 advises C50 of need to protect exposures.

(Radio communications after this point do not clearly reflect the rescue attempts or fire control strategies that were employed.)

APPENDIX B

Area Map

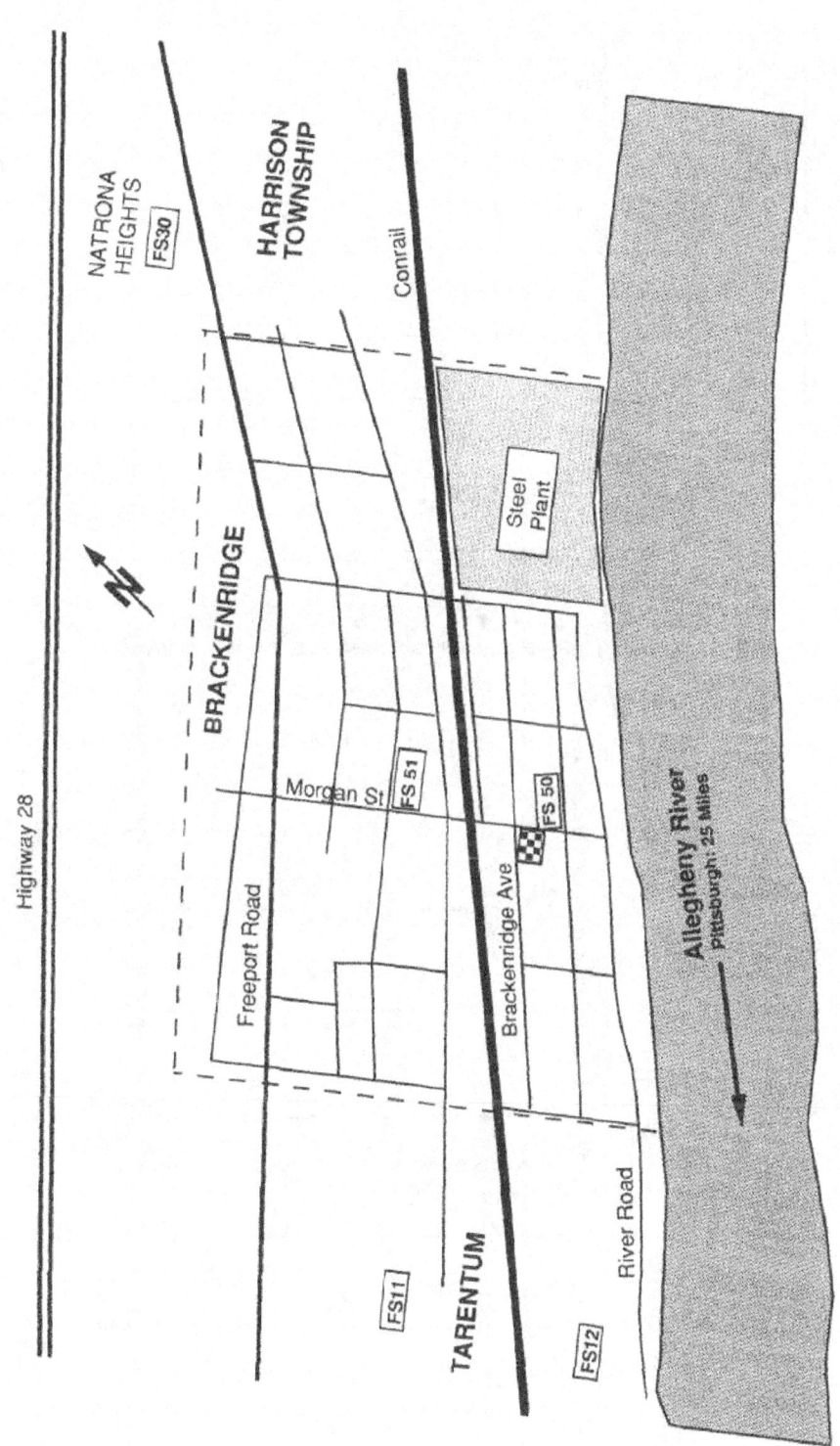

APPENDIX C

Initial Fire Attack and Location of Victims

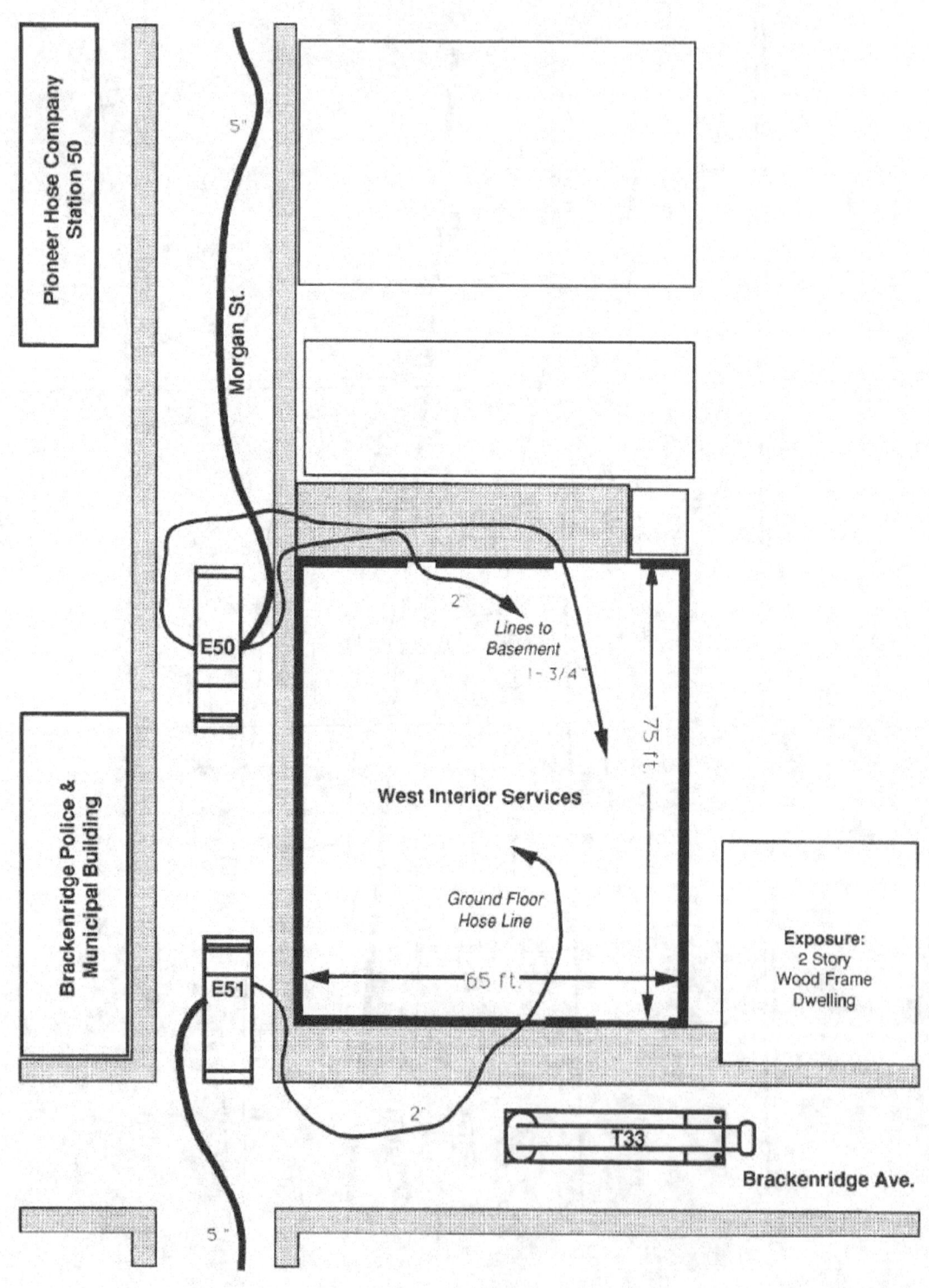

Appendix C (continued)

GROUND FLOOR PLAN

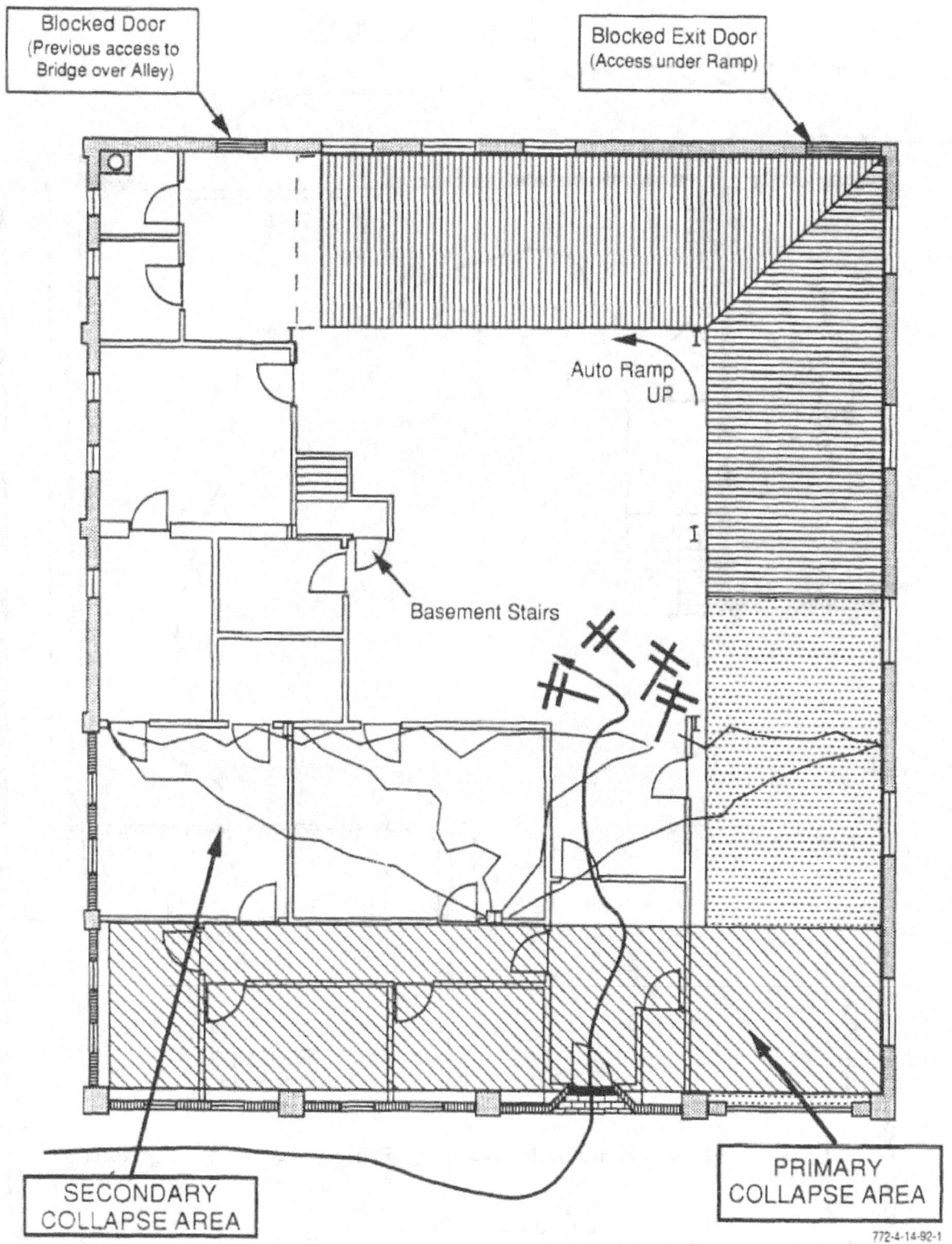

Blocked Door
(Previous access to
Bridge over Alley)

Blocked Exit Door
(Access under Ramp)

Auto Ramp
UP

Basement Stairs

SECONDARY
COLLAPSE AREA

PRIMARY
COLLAPSE AREA

772-4-14-92-1

Appendix C (continued)

BASEMENT PLAN

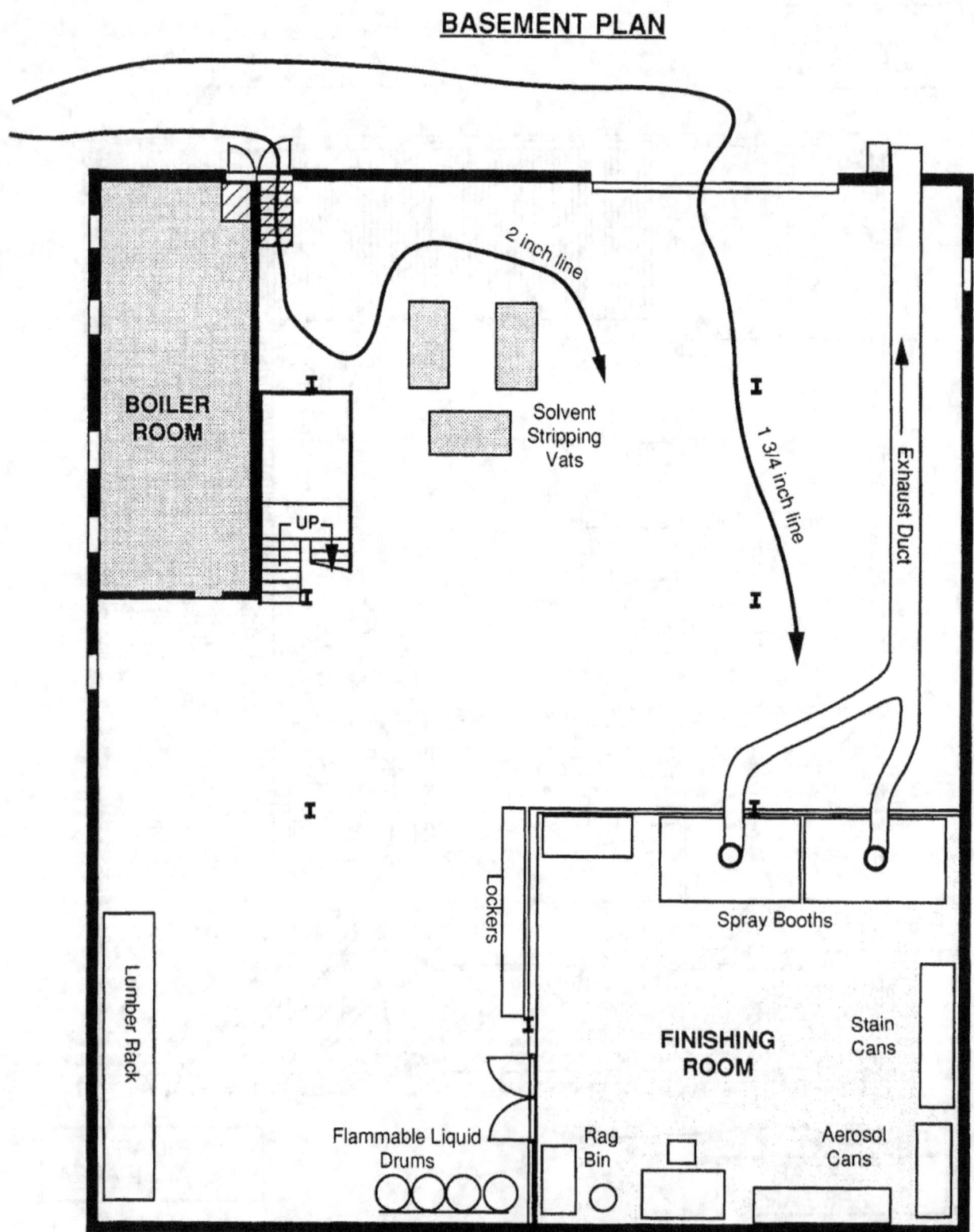

APPENDIX D

Structure Diagrams and Construction Features

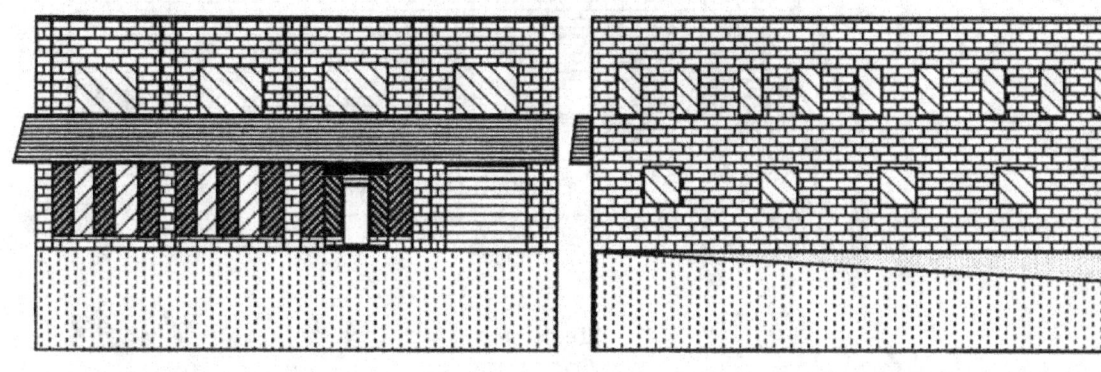

FRONT ELEVATION
Brackenridge Avenue

SIDE ELEVATION

ELEVATIONS

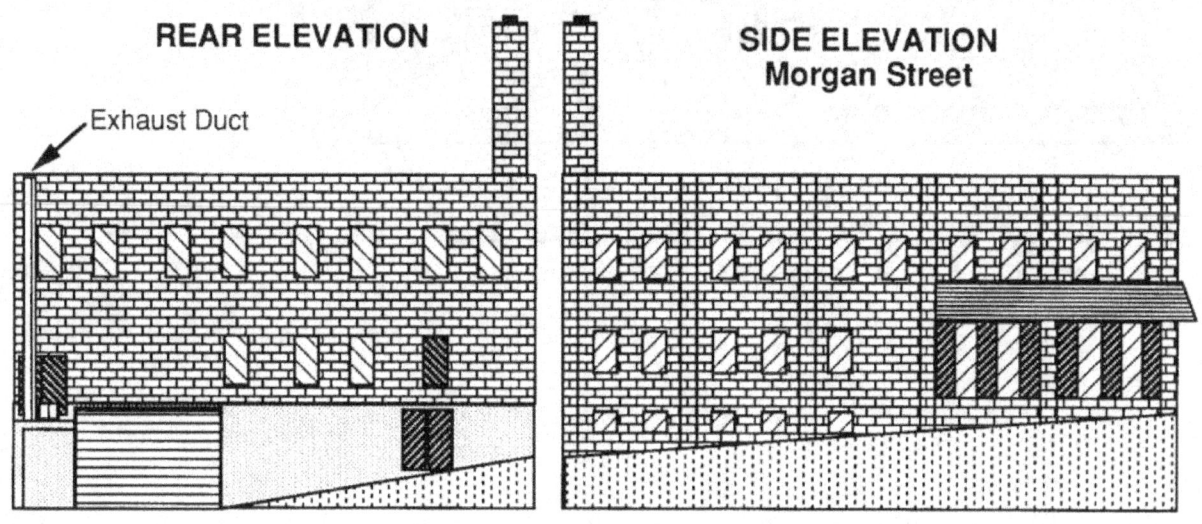

REAR ELEVATION

Exhaust Duct

SIDE ELEVATION
Morgan Street

Appendix D (continued)

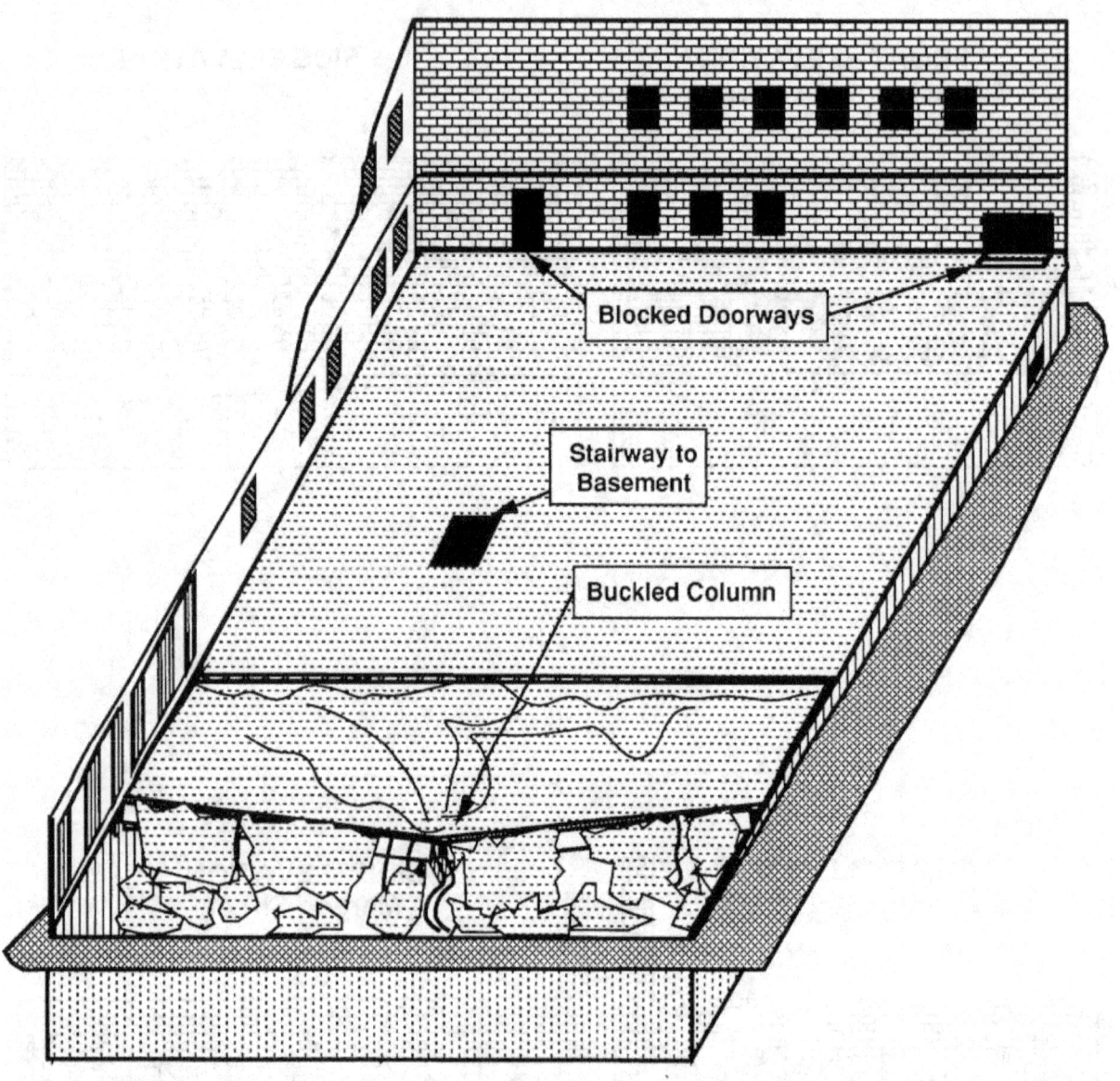

CUT-AWAY VIEW
Showing Floor Collapse

Blocked Doorways

Stairway to Basement

Buckled Column

Appendix D (continued)

CUT-AWAY VIEW
Showing Basement Columns and Beams

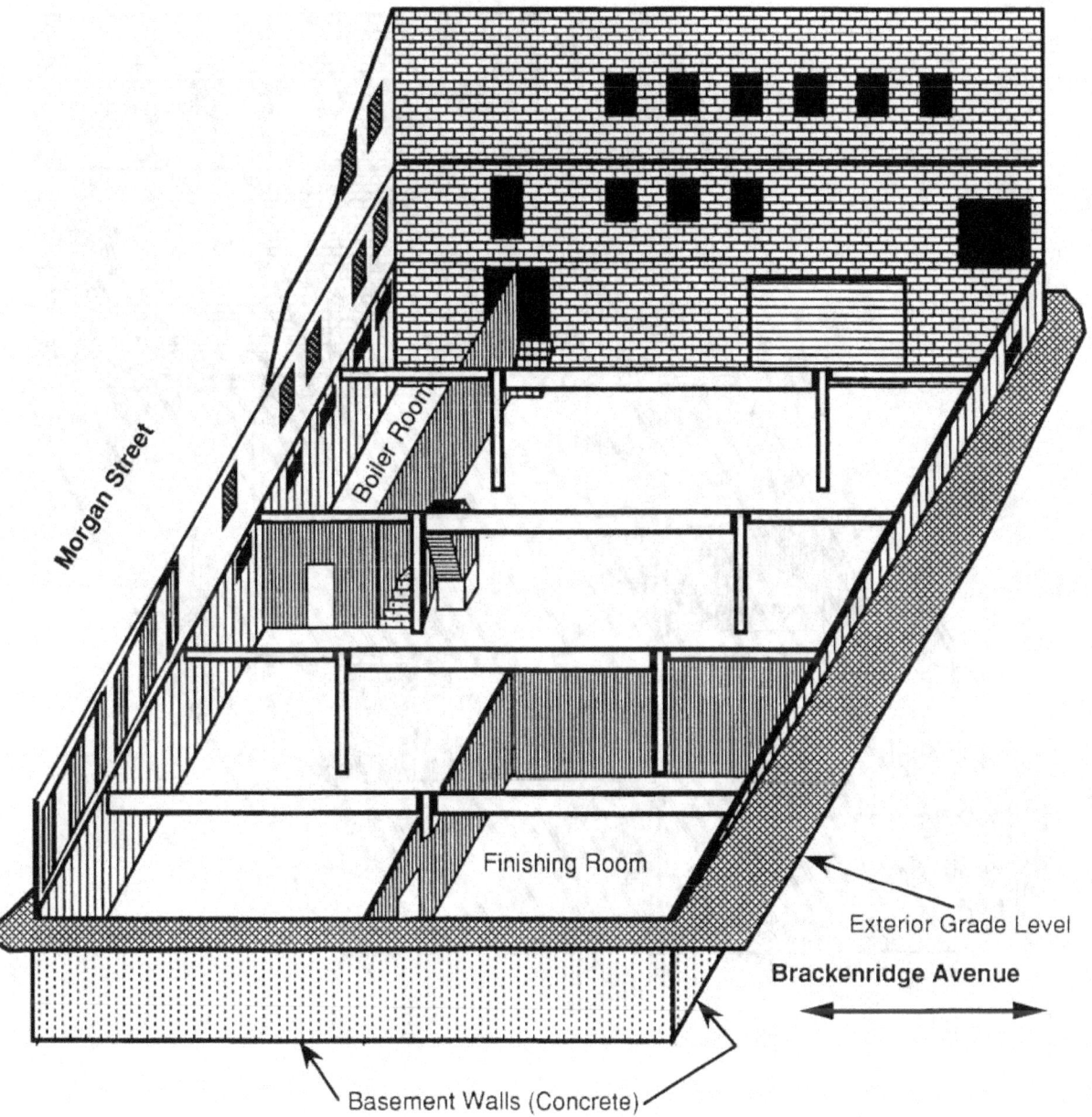

Appendix D (continued)

CUT-AWAY VIEW
Showing Ground Floor Joists

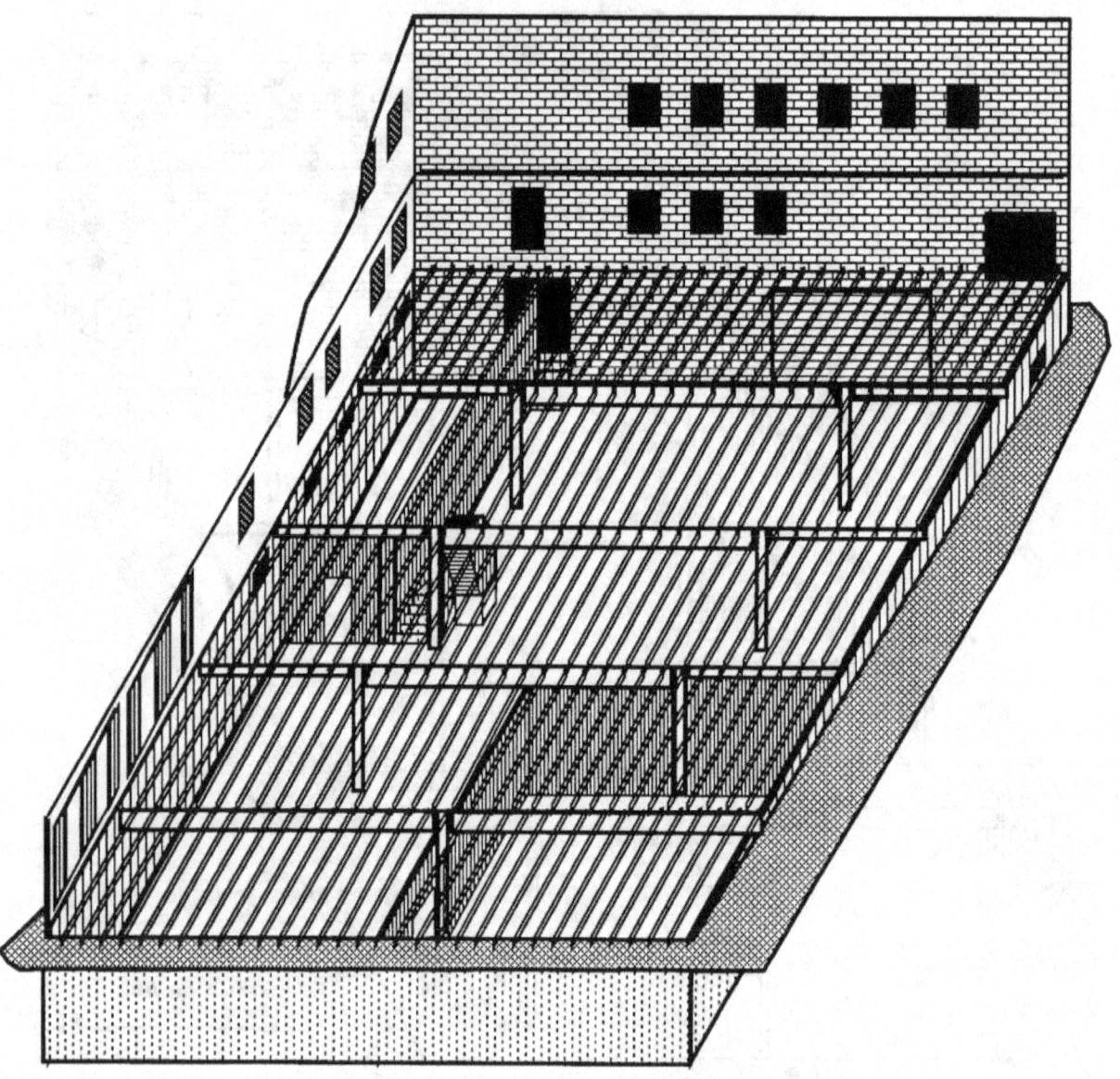

APPENDIX E

List of Fire Companies

PLEASE NOTE: LIST OF FIRE COMPANIES MISSING
FROM TECHNICAL REPORT.

APPENDIX F

Photographs

Copyright 1991, Valley News Dispatch, Tarenton, PA, William T. Larkin

At the height of the fire; firefighters have pulled back.

Appendix F (continued)

Aerial view of fire building after the fire.

Appendix F (continued)

Photo by J. Gordon Routley

Looking toward Brackenridge Ave. from ground floor, near location where bodies were found. Debris in foreground is from first and second floors and from roof structure. Bent-over ends of floor joists are visible at the point of attachment to the basement wall.

Appendix F (continued)

Photo by J. Gordon Routley

**Point of rigid support for floor joists at front of building. Floor joists rested
on concrete basement wall and were held in place by brickwork.
(Steel girder is from roof structure.)**

Appendix F (continued)

View from front (Brackenridge Ave.) looking toward rear of ground floor. Ramp to upper level is visible at rear wall. Monitor nozzle is placed at location of roll-up door.

Photo by J. Gordon Routley

Appendix F (continued)

Photo by J. Gordon Routley

Ground floor area, showing unprotected steel column and ramp to second floor.

Appendix F (continued)

Photo by J. Gordon Routley

Intersection of Brackenridge Ave. and Morgan St., showing four wreaths placed in debris to honor dead firefighters. One of the large roof girders rests on top of the debris. Front portion of the structure was demolished during recovery operations. Note the close exposure between the fire building and the adjacent dwelling. Upper level of the dwelling was damaged by the fire, and one of the steel window frames from the fire building rests against the adjacent structure.

Appendix F (continued)

Photo by J. Gordon Routley

View from sidewalk in front of building (Brackenridge Ave.) looking into basement. Note thin concrete slab, supported by steel joists resting on steel beams.

Appendix F (continued)

Photo by J. Gordon Routley

Partial view of exhaust duct for spray booth air discharges. Rear of metal spray booths and remnants of enclosure wall are visible.

Appendix F (continued)

Photo by J. Gordon Routley

Connection between buckled center column and main beam. Note floor section hanging from opposite side of beam.

Appendix F (continued)

Photo by J. Gordon Routley

Detail of buckled center column, showing damaged connection of horizontal beam.

Appendix F (continued)

Photo by J. Gordon Routley

Basement steel beams, encased in concrete, on each side of unprotected column.

Appendix F (continued)

Photo by J. Gordon Routley

View of unprotected steel supporting second floor. Sag in joists resulted from exposure to the heat of the fire, although this area was not heavily involved in flames.

Appendix F (continued)

Photo by J. Gordon Routley

Stairs from rear entrance door into basement. This entrance was used for the first 2-inch attack line entering the basement.

Appendix F (continued)

Photo by J. Gordon Routley

Roll-up door at the rear of the fire building has been cut with circular saws to allow crews to enter the basement. Initial entry by Eureka crew was through the bottom panels – larger opening was made later.

Appendix F (continued)

Rear driveway. Note beams spanning driveway, which used to support bridge between buildings.

Appendix F (continued)

Photo by J. Gordon Routley

Looking from Pioneer Hose Company Station 50 toward fire building. Entrance to rear driveway (access to basement) is between buildings. Bridge once spanned driveway from second floor of building on left to the fire building.